BLASTOFF!

MERCURY

BLASTOFF!

MERCURY

by Tanya Lee Stone

BENCHMARK BOOKS

MARSHALL CAVENDISH

NEW YORK

In memory of my remarkable uncle, William Gaylor Stone.

With special thanks to Professor Jerry LaSala, University of Maine,
for his careful review of the manuscript.

Benchmark Books
Marshall Cavendish
99 White Plains Road
Tarrytown, NY 10591
www.marshallcavendish.com

Library of Congress Cataloging-in-Publication Data
Stone, Tanya Lee.
Mercury / Tanya Lee Stone.
p. cm. — (Blastoff!)
Includes bibliographical references and index.
Summary: Traces the development of our knowledge of Mercury's orbit, surface features,
and atmosphere from the third century B.C. to the present.
ISBN 0-7614-1403-7
1. Mercury (Planet)—Juvenile literature. [1. Mercury (Planet) 2. Outer space-Exploration.]
1. Title. II. Series.

QB611 .S76 2002   523.41-dc21   20001043915

Printed in Italy
1 3 5 6 4 2
Photo research by Anne Burns Images

Cover Photo: *Photo Researchers, Inc.*: A. Gragera, Latin Stock/Science Photo Library

The photographs in this book are used by permission and through the courtesy of:
Photo Researchers, Inc.: 7: Frank Zullo: 8; David A. Hardy/Science Photo Library: 11;
LOC/Science Source: 12, 13, 15, 16, 17; Science Photo Library: 19; David Parker/Science Photo
Library: 31; Dr. Michael Ledlow/Science Photo Library: 34; Lynette Cook/Science Photo
Library: 46; NASA: 21, 22, 27, 37, 38, 43, 45; *Julian Baum*: 29, 30, 33, 40–41, 55; Astronomical
Society of the Pacific: 42; *The Johns Hopkins University Applied Physics Lab*: 49, 52.

Cover: Artwork representing the planet Mercury, the planet closest to the Sun.

CONTENTS

1

EARLY DISCOVERIES OF MERCURY

Thousands of years before telescopes existed, people watched the night sky. As they looked at the heavens, some wondered about a few stars that seemed to move in and around the others. They seemed different than the other stars. They came to be called wandering stars and were named planets, from the Greek word for wanderer.

These objects in the sky fascinated the earliest astronomers. The Sumerians settled in Mesopotamia, in what is now Iraq, in 3000 B.C. The Sumerians worshiped many gods and believed the moving objects in the sky were gods' homes. Mercury was one of the planets the Sumerians observed. The Babylonian people lived in the same region and used clay tablets to document the movements of the Moon, as well as Mercury and other planets. These tablets are dated between 1700 and 1680 B.C.

In 729 B.C. the Assyrians conquered the Babylonians, after which their empire reached from Egypt to the Persian Gulf. The Assyrians further developed our knowledge of the heavens. They mapped stars and constellations and created a reliable calendar based upon the movements of the Moon and stars. The planet we call Mercury today was named Nebo, or Nabu, by the Assyrians. Nebo was their god of writing.

Because Mercury is so close to the Sun, it can only be viewed from Earth before sunrise or after sunset. This is Mercury seen from the Swift Trail in the Pinaleno Mountains of Arizona at dawn.

Artwork of Mercury and the Sun, as they would appear to a spacecraft orbiting the planet. Mercury is so close to the Sun—and so small—that it is almost impossible to see from Earth.

In 265 B.C. a Greek astronomer named Timocharis studied Mercury and made the first dated observations of the planet. He incorrectly believed that he saw two planets—one that appeared before sunrise and one after sunset. He named them Apollo and Hermes. Apollo was the Greek god of the Sun and light, while Hermes was the Greek messenger of the gods and a guide for lost souls and wanderers.

The Romans named the planet Mercury for their mythological fleet-footed messenger of the gods because Mercury moves quickly across the sky. In fact, Mercury orbits the Sun faster than any other planet. Its average speed is 107,000 miles (172,195 km) per hour. In comparison Earth travels at a speed of 66,490 miles (107,000 km) per hour. And Pluto orbits the Sun the slowest, at the rate of 10,580 miles (17,026 km) per hour.

Mercury is the planet closest to the Sun. It is also the second smallest planet. It is Mercury's small size and closeness to the Sun, rather than its swiftness that makes Mercury so hard to see from Earth.

THE TELESCOPE AIDS UNDERSTANDING

The invention of the telescope in the early 1600s began centuries of detailed observations that would bring us closer to understanding the mysteries of our Solar System. In 1610 Italian astronomer Galileo Galilei became the first known person to view Mercury through a telescope. As telescopes improved, another Italian astronomer, Giovanni Zupus, studied the planet more closely. In 1639 Zupus discovered that Mercury has phases. This strengthened the case for what astronomers already believed—that Earth circles the Sun. Just two years later, in 1641, German astronomer Johann Franz Encke was the first person to calculate the mass of Mercury.

Nearly 250 years passed before any more significant information about Mercury was gathered. In 1889 the world got its first look at a map of the planet. Italian astronomer Giovanni Schiaparelli drew the

The Mystery Planet

In 1859 a French astronomer and mathematician named Urbain Le Verrier gave a lecture describing a new planet positioned between Mercury and the Sun. He named it Vulcan, after the Roman god of fire. He believed this planet explained why calculations of Mercury's orbit were always slightly wrong. He reasoned that something was pulling Mercury off course, and announced that the culprit was the yet-to-be discovered planet Vulcan. He even calculated Vulcan's orbit around the Sun. Astronomers watched for Vulcan on the days Le Verrier predicted it would be observable. There were a few so-called "sightings" over the next several years, but no evidence was ever found that proved Vulcan's existence.

A respected amateur astronomer from France, Edmund Lescarbault, reported having seen a round black spot on the Sun on March 26, 1859, that looked like a planet traveling across it. Le Verrier investigated the observation and computed an orbit from it. The orbit was estimated to be 19 days 7 hours. The diameter was much smaller than Mercury's and its mass was one-seventeenth of Mercury's mass. This was too small to account for the deviations of Mercury's orbit, but might have been a sunspot or an asteroid. The planet Vulcan was never discovered—because it did not exist. But something still had to explain what seemed to be warping Mercury's orbit.

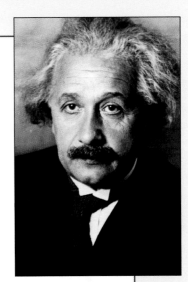

In 1915 physicist Albert Einstein published his General Theory of Relativity. Einstein's theory changed the way we see the world. Because of Isaac Newton's laws of universal gravitation, astronomers already knew that planetary orbits are not perfectly circular. Instead, they are elliptical. But Einstein revealed that the ellipse *itself* also rotates, causing a wobbling or warping of the orbit. This is called a precession. Over time the axis (the imaginary line running from the north to south pole of a planet) of Mercury's orbit shifts slightly. Imagine spinning a top on a table. If you leave it alone it runs its course unchanged. But if you nudge it, changing its axis, the orbit changes position.

Einstein was excited by the results of his theory. He wrote, "Imagine my joy . . . that my equations account for the motion of Mercury." To this day, Mercury's particular type of elliptical orbit supports the General Theory of Relativity.

Albert Einstein (1879–1955), famous for his General Theory of Relativity, discovered that Mercury's ellipse rotates. This led to our present-day understanding of the odd way in which the planet rotates.

This art depicts Italian astronomer Galileo Galilei (1564–1642) demonstrating his telescope. It was the first to be used for astronomical observations. In 1610 Galileo viewed Mercury, making him the first known person to see the planet through a telescope.

map based on his observations. He had become well known in 1877 for his map of the planet Mars on which he labeled certain features he called canals. Schiaparelli's map of Mercury contained similar canallike markings. The popular view was that Schiaparelli believed these were manmade canals; in fact, that interpretation was made by others, who translated his work incorrectly.

THE FIRST TELESCOPES

Telescopes began to be used in the early part of the seventeenth century. Although there are some scientists who believe that there were earlier telescopes, Hans Lippershey is generally credited with building the first one in 1608.

An historical engraving of Hans Lippershey (1570–1619), the Dutch optician who is thought to have invented the first telescope in 1608. Some historians believe that Lippershey's children were playing with a couple of lenses (center right) and found that a distant object looked bigger and closer than it did when they looked at it without the lenses. Lippershey's first telescope magnified objects three times their natural size.

Galileo Galilei (and other craftsmen and scientists) quickly built a telescope based on Lippershey's design that was able to magnify objects by three times. Galileo then improved upon his design several more times over the next year until he built a telescope that could magnify objects by twenty times. It was through this telescope that Galileo made his first discoveries.

There has been a fair amount of controversy over the issue of who truly invented the telescope. Some historians believe that Galileo heard of Lippershey's design and quickly built one to present to officials as his own invention. Others believe that Galileo always credited "the Dutchman" with the idea. There is even a story supported by historians that Lippershey himself got the idea from his two children. They were playing in his workshop with two lenses together and saw an object in the distance that appeared larger through the lenses than it did when they just used their eyes.

The early telescopes were somewhat similar to modern binoculars. They consisted of two tubes connected so that the overall length of the telescope tube could be adjusted to help bring an object into focus. (Think about inserting one paper towel tube inside another to make one long tube.) At each end of the tube was a lens. The lens farthest from the viewer's eye was a curved convex lens. It was used to collect and bend the light emitted from a distant object and bring it into focus. The second lens, placed at the eyepiece of the telescope, was a concave lens that again bent the light, making the image appear more normal to the viewer. The design was later improved by using a second or convex lens as the eyepiece. Although it turned everything upside down to the viewer, the new design greatly improved the quality of the images seen through the telescopes. This type of instrument became known as an astronomical telescope.

The early telescopes were only about 6 feet (1.5 m) long, but in 1656, Dutch astronomer and physicist Christiaan Huygens (1629–1695) made one that was 23 feet (7 m) long.

Many of the earliest telescopes were only 5 or 6 feet long. But by the middle of the 1600s, telescopes grew longer. In 1656 Christiaan Huygens made one that was 23 feet (7 m) long. As telescopes developed, magnification increased and the quality and positions of the lenses improved. Just imagine what Galileo might be doing if he were alive today, pondering the uses of the many different types of telescopes that exist now.

A computer composite image of the nine planets of the Solar System. The planets are (bottom left to top right): Pluto, Neptune, Uranus, Saturn, Jupiter, Mars, Earth, Venus, and Mercury. A representation of the Sun is at top right.

A few years later Greek astronomer Eugene Antoniadi went to work in France. From 1914 to 1929 he made detailed observations of Mercury using a more powerful telescope than the one Schiaparelli had used. Antoniadi then created a new map of Mercury that was

In 1889 Italian astronomer Giovanni Schiaparelli drew a map of Mercury based on his observations of the planet.

used for 40 years. He proved that the canals were an optical illusion. Schiaparelli's Martian canals were also proven incorrect at a later date.

After Antoniadi, no major discoveries were made about the small planet for many years. So, even as late as the 1960s, little was known about Mercury. It was hard to be sure if the calculations of the early astronomers were accurate. But new technology and innovations were just around the corner. People were about to get a whole new view of the planet closest to the Sun.

2

A Closer Look

The first real breakthrough in understanding more about Mercury came in 1965. A new telescope called a radio telescope had been built two years earlier near Arecibo, Puerto Rico. This telescope works by beaming radar signals at a heavenly object and working with the signals reflected back by the object. A huge wire-mesh dish receives the reflected signals and computers are used to look at the data and make sense of it.

With the radar technology, astronomers were able to collect much more accurate information on Mercury's exact location and distance from Earth. Even more importantly, the radio telescope helped astronomers determine just how quickly Mercury was spinning in space—something that had been disputed by astronomers for centuries.

Until 1965 it was generally believed that Mercury rotated on its axis once every 88 Earth days. (A planet's rotation is the amount of time it takes to spin completely around on its axis. This is equal to a planet's day.) Eighty-eight days is the amount of time it takes Mercury to make one revolution around the Sun (a planet's year). Because astronomers thought that the rate of rotation and the time it took for one revolution were both 88 Earth days, they reasoned that the same

This aerial view of the Arecibo radio telescope shows the Gregorian subreflector system that was installed in 1997. The largest radio system in the world, the Arecibo is situated in a natural crater in the mountains of Puerto Rico.

side of the planet was always facing the Sun. Therefore, the theory before 1965 was that one side of Mercury was in constant darkness and the other side was in continuous light.

But that theory was wrong. The information gathered by the Arecibo telescope changed what we know today. Astronomers learned that Mercury actually rotates on its axis once every 59 Earth days. This means that a Mercury day (59 Earth days) is not much shorter than a Mercury year (88 Earth days). And because Mercury rotates slowly in the same direction as it revolves very quickly around the Sun, daytime and nighttime on Mercury each last a long time.

MERCURY'S FIRST CLOSE-UP—*MARINER 10*

The Arecibo telescope helped astronomers enormously, but more could be learned about Mercury by actually getting closer to the planet. This was made possible on November 3, 1973, by the United States when the National Aeronautics and Space Administration (NASA) launched *Mariner 10*.

Earlier *Mariner* space probes had already explored Mars and Venus. *Mariner 10* was different because it was going to venture so near the Sun. To help protect it from the intensity of the Sun, a large sunshade was added to the design. Thermal blankets that control heat were also used to wrap the top and bottom of the probe. The small, eight-sided craft weighed 1,164 pounds (528 kg).

Mariner 10 marked the first time two planets were studied by one spacecraft during the same mission. It was also the first time that scientists had discovered a way to use solar wind to help a craft travel through space. (Solar wind is a stream of charged particles given off by the Sun that travels throughout the Solar System.) When *Mariner 10* ran low on fuel, solar panels were used to catch the wind, much like the sails on a boat.

Mariner 10 reached its initial destination—Venus—on February 5, 1974.

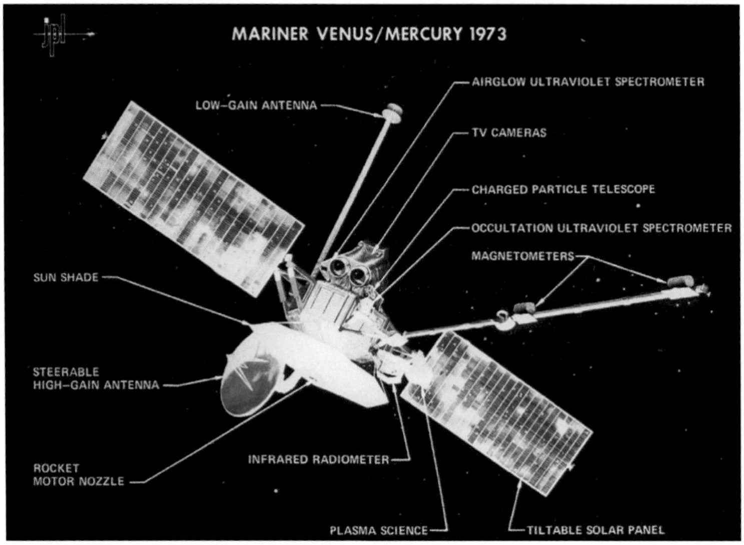
MARINER VENUS/MERCURY 1973
AIRGLOW ULTRAVIOLET SPECTROMETER
LOW-GAIN ANTENNA
TV CAMERAS
CHARGED PARTICLE TELESCOPE
OCCULTATION ULTRAVIOLET SPECTROMETER
MAGNETOMETERS
SUN SHADE
STEERABLE
HIGH-GAIN ANTENNA
INFRARED RADIOMETER
ROCKET
MOTOR NOZZLE
PLASMA SCIENCE
TILTABLE SOLAR PANEL

An artist's rendering of the spacecraft that flew by Venus and Mercury, with all of its parts labeled, before it was launched into space as Mariner 10.

The probe made history again by becoming the first craft to use the gravitational pull of one planet to bend its course and send it speeding toward another—in this case, Mercury. NASA's idea was a success.

It was thrilling to see that this innovative method worked beautifully. During the course of the mission, *Mariner 10* was catapulted past Mercury in an orbit around the Sun and was able to make two more passes before its work was done. (This gravitational "slingshot" technique also was crucial to the successes of the later *Pioneer* and *Voyager* missions to the outer planets.) This gave scientists three different sets of images taken by the space probe each time it flew by Mercury.

The first flyby (the flight of a spacecraft past a celestial body) was on March 29, 1974. The probe was only 437 miles (703 km) from Mercury when it passed by. The second flyby was September 21, 1974, at a distance of 29,870 miles (48,069 km). The last time *Mariner 10* photographed Mercury was on March 16, 1975, from 203 miles (327 km) away.

The **Mariner** *spacecraft weighed about 1,100 pounds (498 kg), including 170 pounds (77 kg) of scientific equipment. It was launched from what is now called Cape Canaveral, Florida, in 1973.*

Each time, *Mariner 10* transmitted images back to Earth.

The mission results were overwhelming. Never before had we been able to see close-up images of this planet. The *Mariner 10* mission resulted in more than ten thousand images of Mercury. This allowed scientists to accurately map more than half (57 percent) of the planet that was lit by the Sun each time the probe flew by.

THE EXPERIMENTS

In addition to the photographic equipment it carried, *Mariner 10* also had several instruments onboard to measure different characteristics of the planet. In all, seven experiments were conducted.

The Television Photography experiment had very specific goals. Photographs were taken that helped scientists map the major physical features of the planet. Craters and long expanses of cliffs and plains were all photographed by *Mariner 10*. This experiment also set out to search for satellites of Mercury and gather more information on the shape of the planet. *Mariner 10* discovered that Mercury has a shape that is rounder and more spherical than Earth's. Combinations of telescopes attached to both wide-angle and narrow-angle television cameras allowed *Mariner 10* to take images from different angles and perspectives. No satellites of Mercury were found.

Another experiment, called Celestial Mechanics and Radio Science, used radio equipment to learn more about Mercury's mass. This allowed scientists to accurately confirm the density of Mercury.

Mariner 10 also performed an experiment to measure the plasma that comes from the Sun in the solar wind. Three detectors were used —two facing toward the Sun and one facing away from the Sun. The detectors measured protons and electrons to learn more about the interaction between the solar wind and Mercury. The experiment also studied the direction and speed of the solar wind.

What scientists expected to find—since they had not yet discovered

THE ARECIBO TELESCOPE

Radio telescopes work by studying the long wavelength electromagnetic radiation that is given off by objects in space. The largest stationary telescope in the world is the Arecibo radio telescope near Arecibo, Puerto Rico. The opening ceremony was held on November 1, 1963. By April 7, 1964, the telescope had picked up signals from Mercury. The following year Arecibo celebrated one of its first major achievements. Scientists were able to accurately calculate the number of Earth days it takes Mercury to make one full rotation. Instead of 88 days, researchers at Arecibo correctly determined the rotation to be 59 days. The world's most sensitive radio telescope had started with an important discovery.

The Arecibo telescope is an engineering marvel. The enormous bowl-shaped reflecting dish is made up of almost

that Mercury has a magnetic field—was evidence that the solar wind reaches the surface of Mercury. (Earth's magnetic field deflects the solar wind and it does not reach Earth's surface.) But instead, scientists were amazed to see data showing that the path of the solar wind is disturbed around Mercury. This added to the growing evidence of a magnetic field around Mercury. Both the solar wind and the magnetic

forty thousand aluminum mesh panels that are supported by a maze of steel cables. The dish itself is 1,000 feet (305 m) across. Thirteen football fields would fit inside it! It was built over a large natural sinkhole to accommodate the dish. Even with the advantage the sinkhole offered, additional rock had to be blasted to make room for the telescope.

A platform hangs 450 feet (137 m) above the dish from a triangular support system of three concrete towers and eighteen cables. All of the technology that sends, receives, and focuses signals from space onto the dish is suspended from this central platform. The equipment includes radar transmitters, radio receivers, and a variety of antennas. Nearly 40 years after the Arecibo telescope first began to operate, it is still our best tool for detecting faint radio signals from space.

field will be studied further in future missions to Mercury.

The Magnetic Field experiment used instruments called magnetometers to detect and measure Mercury's magnetic field. The sensors were protected from the intense radiation of the Sun by thermal blankets and sunshades. This experiment resulted in the surprising discovery of a weak magnetic field. Scientists had not expected the probe

Mercury's Size

Mercury is the second-smallest planet in the Solar System. At only 3,030 miles (4,876 km) in diameter, eighteen Mercurys could fit inside Earth.

to find a magnetic field because of how slowly the planet rotates.

The purpose of the Extreme Ultraviolet Spectrometer experiment was to study Mercury's atmosphere. This experiment used two instruments that measured ultraviolet light (radiation beyond the violet end of the visible spectrum). One was stationary, while the other could be pointed in different directions. The instruments found evidence of the elements hydrogen, helium, and oxygen in the atmosphere.

Mariner 10 also measured the range of temperatures found on Mercury with the Two-Channel Infrared Radiometer Experiment. An infrared radiometer detects and measures electromagnetic radiation

Distance from the Sun

Mercury orbits closest to the Sun—it is an average of 36 million miles (58 million km) from the Sun. In comparison Earth is the third-closest planet to the Sun, at a distance of 93 million miles (150 million km).

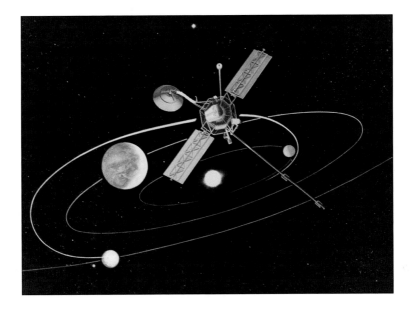

In November 1974 NASA's Mariner 10 *mission flew by the planets Venus and Mercury. This artist's rendition shows the rocket returning visible data to Earth.*

beyond the red end of the electromagnetic spectrum that is not visible to the naked eye. It compared the radiation coming from the planet with the radiation levels coming from black space. This sensitive instrument was mounted on the outside of the probe and protected by thermal blankets that helped shield it from the scorching Sun.

Two telescopes were used in the Energetic Particles experiment. This experiment measured the charged particles around Mercury. Like the Plasma experiment, this experiment also obtained unexpected results while it focused on learning more about how the solar wind interacts with Mercury. It found additional evidence of something around Mercury interrupting the path of the solar wind.

The *Mariner 10* mission ended on March 24, 1975, after completing some final measurements. The probe's controls no longer work and it is out of fuel. But just as the Moon is held in orbit around Earth —captured by its gravitational pull—*Mariner 10* will remain in orbit around the Sun indefinitely.

3

MERCURY'S ENVIRONMENT

Much of what we know about Mercury today is a result of the *Mariner 10* mission. Among the major discoveries, the space probe detected Mercury's atmosphere and its magnetic field. It also helped scientists determine the range of temperatures experienced on the planet closest to the Sun.

A THIN ATMOSPHERE

Mercury does have an atmosphere—but just barely. It is composed of trace amounts of oxygen, hydrogen, sodium, potassium, and helium. Some of these elements come from the solar wind and some come from the materials on the surface. The atmosphere is so thin that it does not hold together in the same way as Earth's atmosphere. Mercury's loosely formed atmosphere is constantly being blasted by the solar wind and escaping into space. For this reason Mercury's atmosphere is not stable. Instead, it is constantly being replenished.

Mercury's extremely thin atmosphere affects the planet in several ways. It offers no protection against constant bombardment from space materials. Asteroids and comets continue to smash into the surface of

Mercury's atmosphere is so loosely formed that it doesn't protect the planet from bombardment by objects in space. Here an artist depicts an asteroid hurtling into the planet.

The atmosphere of Mercury is so thin that it keeps coming apart. It must constantly replenish itself.

Mercury and add to its huge assortment of craters. The low atmospheric pressure also makes Mercury a very silent place. Sound waves do not travel in outer space, where there is no atmospheric pressure. On Mercury the atmospheric pressure is much lower than it is on Earth. That is why sound travels much better on Earth than it does on Mercury. The other eerie aspect of Mercury is its sky. You might imagine that it is always bright on Mercury because it is so close to the Sun. But the opposite is true. Mercury's sky is always black. That is because Mercury's atmosphere is so thin that it doesn't scatter the light. The scattering of the light is what makes our sky blue.

A VAST TEMPERATURE RANGE

The atmosphere of a planet acts as an insulator, working to keep heat from escaping. On Earth, the atmosphere helps keep temperatures steady. Because Mercury lacks any substantial atmosphere, there is nothing to help regulate temperatures on Mercury. In fact, no other planet or satellite in our Solar System experiences such drastic ranges of temperature as Mercury. During the day, Mercury can have temperatures as high as 805 degrees Fahrenheit (429 °C)—that is hot enough to melt lead! In sharp contrast, nighttime temperatures can fall as low as –266 degrees Fahrenheit (–165 °C).

Because Mercury barely has an atmosphere, temperature is impossible to control. Mercury has the most extreme temperature ranges of any planet. This false color temperature map of Mercury shows the extremes. Just below the planet's surface, red indicates the highest temperature of 260 degrees Fahrenheit (126 °C).

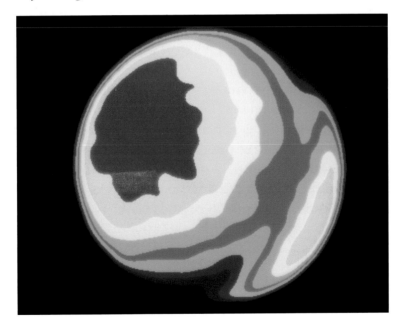

Another factor that affects this enormous span of temperature is the shape of Mercury's orbit, which is extremely elliptical. When Mercury is at its closest to the Sun (perihelion), it is only 29 million miles (47 million km) from the hot star. At its farthest (aphelion), Mercury is 43 million miles (69 million km) from the Sun. In comparison, the average distance between Earth and the Sun varies between 91 million (146 million km) and 94 million (151 million km) miles.

Because of the way Mercury spins on its axis while it revolves around the Sun, two areas of the planet experience the most extreme temperatures. These regions of Mercury are closest to the Sun during perihelion. They receive more than two and a half times the amount of radiation from the Sun as other areas of Mercury. It is easy to imagine why they are called the hot poles.

What Would You Weigh on Mercury?

Mass and weight are commonly confused. But mass and weight are not the same thing. Mass is the amount of material that is in something—an apple, a planet, or your body. Weight is the measure of how much gravity is pulling on a mass. The gravitational field of Mercury is a little more than a third that of Earth's. A 100-pound (45 kg) person on Earth would weigh nearly 38 pounds (17 kg) on Mercury.

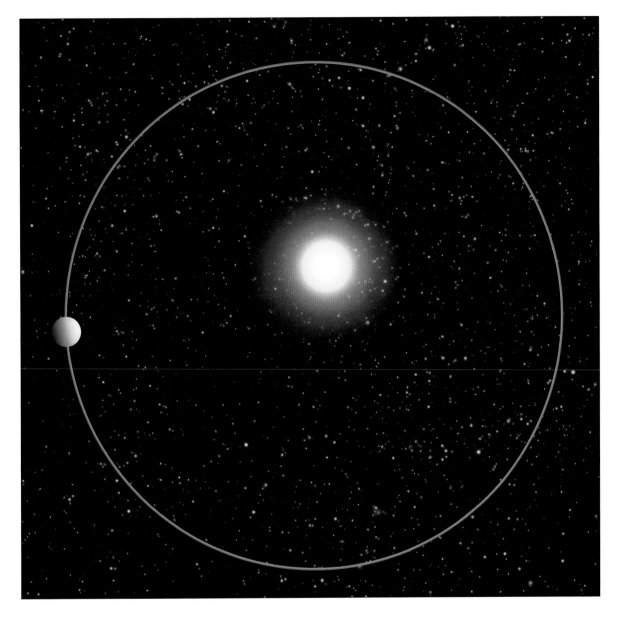

Mercury's orbit is extremely elliptical. Its unusual orbit affects the range of temperature on the planet.

This cutaway illustration shows the interior structure of Mercury. The planet's large central core consists of iron, which accounts for about 75 percent of Mercury's diameter. Surrounding this is a rocky mantle and then a thin crust of sun-baked rock.

A MAGNETIC FIELD AND AN IRON CORE

Before *Mariner 10*, scientists did not think that Mercury had a magnetic field. The discovery the probe made surprised the experts. Mercury does have a magnetic field, although it is very weak—about one hundred times weaker than Earth's. But what exactly is a magnetic field and where does it come from? It is the area in which a magnetic force can be felt. Astronomers think the iron core of a planet acts as an electrical conductor and generates current. This current is what creates the magnetic field of a planet.

Mercury's iron core is believed to be quite large. It extends about three-quarters out from the center of the planet for 2,250 miles (3,621 km). Above the core a rocky mantle and crust are only about 400 miles (644 km) thick. Mercury's massive, heavy iron core is what makes the planet so dense. We still have a lot to learn about Mercury's magnetic field and whether there is liquid molten iron in the core. The only planet that has a greater density than Mercury is Earth. Since its liquid iron core generates Earth's magnetism, it seems probable that Mercury also has liquid iron at its core. A mission scheduled to travel to Mercury in 2004 will tell us more about this elusive planet.

4

MERCURY'S TERRAIN

Scientists were excited by the amount of knowledge that was obtained through the *Mariner 10* mission. Mercury has been described as being similar to Earth's Moon due to its heavily cratered surface and seemingly dead terrain. But aside from the craters and its dusty covering, Mercury has geological features that are quite different from any found on the Moon. For example, Mercury has enormous cliffs that look like wrinkles on the surface and extend for hundreds of miles. There are also large areas of smooth plains unlike any terrain on the Moon.

IMPACT CRATERS

Like other planets in our Solar System, Mercury has been bombarded with countless asteroids. (Once an asteroid lands it is called a meteorite.) The result of this space material crashing into Mercury is the many craters that pepper the surface of the planet. Still other impressions were made by rocks thrown up from the impact of the blasts that then fell back down onto the surface. Many of the craters on Mercury have been there for billions of years. Comets crashing into

Mercury has been bombarded with so many asteroids that it is covered with craters. A new crater is shown in this picture of Mercury taken by the Mariner 10 *spacecraft. The crater is 7.5 miles (12 km) across.*

Mercury's Caloris Basin is so big it could easily hold the state of Texas inside. The basin is 800 miles (1,300 km) in diameter and bounded by mountains which rise higher than a mile up (2 km).

the planet at high speeds have caused other, more recent craters. The number of bowl-shaped craters on the surface accounts for some of the comparisons made between Mercury and the Moon.

Mercury's Caloris Basin is one of the largest craters in the entire Solar System. Some think it looks like a giant bull's-eye on the surface of Mercury. The Caloris Basin is about 807 miles (1,299 km) in diameter. The state of Texas could fit inside it! It was created by the impact of an enormous asteroid smashing into the planet—most likely in Mercury's early history. Scientists estimate this event took place 3,850 million years ago. The impact was so great that it sent shock waves traveling through the planet to the opposite side. The tremors pushed up hills on the other side of Mercury.

Many rings that were created upon impact rim the outer edges of the Caloris Basin. They look similar to the rings that flow outward when you toss a rock into the water. Some of these mountainous rings reach heights of 6,500 feet (1,981 m). If you were able to peer down into the Caloris Basin, you would see plains that are interrupted by wrinkled ridges and fractures in the surface. Some of the ridges are up to 185 miles (298 km) long but less than 1,000 feet (305 m) high. Many smaller craters, formed at a later time, also dot the floor of the Caloris Basin.

Another interesting crater found on Mercury is the Kuiper Crater. This large impact crater is about 37 miles (60 km) across. It was named for the Dutch-American astronomer Gerard P. Kuiper. Kuiper was part of the team that developed the photographic technology for *Mariner 10*, but he died a few months before the mission reached Mercury. The Kuiper Crater is unique because it reflects sunlight more strongly than any other known Mercurian feature.

SMOOTH PLAINS AND VOLCANOES

Expanses of smooth plains are common on Mercury, just as they are on Mars, Venus, and Earth. Plains develop in many different ways.

One way is for material to fall down onto an area of a planet. This
material could come from volcanic ash or the debris generated during
the creation of an impact crater. An area becomes blanketed, smooth-
ing out any rough terrain that had been underneath it.

Plains can also be the result of lava flows. Lava from volcanic
activity accounts for many of the smooth plains found on Mars,

Mercury has large expanses of smooth plains. Because of the Mariner 10 *mission, scientists now believe that there were once active volcanoes on Mercury.*

Venus, and Earth. Knowledge of this sparked the question of whether or not there had ever been active volcanoes on Mercury.

This question has begun to be answered using *Mariner 10* data. Although the data from *Mariner 10* was originally collected in 1974

GERARD KUIPER

Astronomer Gerard P. Kuiper, known as the father of modern planetary science, was part of the Mariner 10 *team.*

Gerard P. Kuiper was born in the Netherlands on December 7, 1905. He moved to the United States in 1933, after earning an advanced degree in astronomy and physics. He began working at the Lick Observatory of the University of California that same year. Four years later, Kuiper became a U.S. citizen. He later worked for Harvard University, the U.S. Office of Scientific Research and Development, and the University of Chicago. In 1960 Kuiper founded the Lunar and Planetary Laboratory (LPL) at the University of Arizona in Tucson. Researchers at LPL study how our planetary system was formed and how it has changed over time.

Kuiper is known as the father of modern planetary science because of his many important achievements. In 1944 he determined that the atmosphere of Saturn's satellite Titan contains methane and ammonia, making it similar to Earth's atmosphere. In 1948 Kuiper determined that the rings of Saturn are made up of ice particles. That same year, he discovered the fifth moon of Uranus and named it Miranda. The following year he discovered Nereid, Neptune's second moon.

In the 1950s, Kuiper further advanced the field of astronomy. He computed Pluto's rotation (6.4 days), observed volcanic eruptions on Jupiter, and helped us understand the polar caps on Mars. In the 1960s Gerard

Kuiper was an important asset to the space program. He helped choose a safe landing site for the return to Earth of *Apollo 11*, which had landed on the Moon on July 20, 1969.

One of Kuiper's major contributions to the study of our planetary system stemmed from his belief that we could never learn enough about the stars and planets using only ground-based telescopes. He knew that objects in the sky had to be viewed from a much higher vantage point. Kuiper was influential in the effort to have NASA operate a telescope from the air.

In 1967 NASA equipped an airplane with an infrared telescope. Kuiper used this technology with great success, which led to NASA supporting a second, more sophisticated project. However, it was not finished until two years after he died. The Kuiper Airborne Observatory (KAO) was dedicated to him in May 1975. The KAO continued to operate until October 1995, when it was retired to make way for another, more advanced project. This new project, the Stratospheric Observatory for Infrared Astronomy (SOFIA), has been in operation since 2000. Kuiper's passion for learning about our Universe may have contributed more to the study of our planetary system than the work of any other modern astronomer.

The Mariner 10 *team proposed the name "Kuiper" for this bright crater at the rim of an older crater. Kuiper died while the spacecraft was en route to Venus and Mercury.*

and 1975, some pieces were reanalyzed in 1997. Mark Robinson was a researcher working for the U.S. Geological Survey in Flagstaff, Arizona. His collaborator, Paul Lucey, was based at the University of Hawaii. Together they published their results in a scientific paper that appeared in the journal *Science* in 1997. Through their studies, Mark Robinson and Paul Lucey discovered that some of the smooth plains on Mercury may have been formed by lava that flowed into the area and later cooled. It now seems likely that the planet experienced volcanic eruptions in its early history.

COMPRESSION CLIFFS

There are many cliffs on the surface of Mercury that run from 100 to 350 miles (161 to 563 km) in length and are several miles high. They look like gigantic wrinkles in the planet's crust and are called escarpments, or scarps. Mercury's scarps run right through craters and other surface features.

Scientists believe that these cliffs were formed after Mercury experienced a dramatic cooling of its iron core. This caused the entire planet to compress, or shrink, a little more than half a mile (1 km) in radius. (The radius is an imaginary line that goes from the center to the outside of a circle or sphere.) These surface features resemble the marks you might find on a leathery, crinkled piece of fruit left out to dry.

Several of Mercury's scarps have been named after explorers' ships. The Discovery Scarp is named for one of Captain Cook's ships, which sailed the Pacific Ocean in the late 1700s. This cliff is 340 miles (547 km) long and cuts through two large craters. The Santa Maria Scarp took its name from one of the three ships on Christopher Columbus's famous expedition to America in 1492. The Victoria Scarp is named after the first ship to sail around the world, an expedition that was led by explorer Ferdinand Magellan.

A cratered area near Mercury's south pole was photographed by Mariner 10 *during its second flyby in 1974. The plains between the craters are crossed by ridges and scarps.*

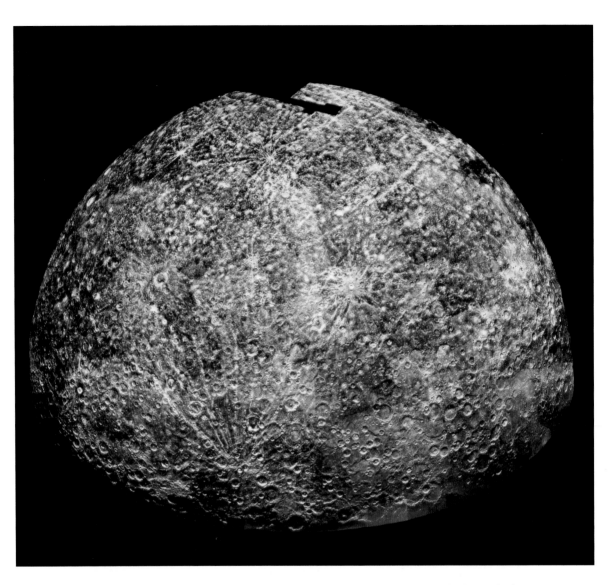

Mercury's south pole is believed to have ice on its surface. This photograph shows the heavily cratered southern hemisphere.

ICE AT THE POLES

Can you imagine that the planet closest to the Sun might actually have ice on the surface? Remember that a day on Mercury, from sunrise to sunrise, is equal to about 176 Earth days—a long time to have constant exposure to the Sun. But information collected in the early 1990s by the Arecibo, Goldstone, and Very Large Array (VLA) Earth-based radar technology suggested that ice did indeed exist on Mercury.

The evidence for ice exists in the form of areas near Mercury's north and south poles that appear to be very bright. There are many craters in these areas that could exist in permanent shadow, allowing for the possibility of ice. A large crater at the south pole, named Chao Meng-Fu, marks the spot of one of the largest of these water-ice areas. In all, about twenty bright reflective areas have been detected by radar. This water could have come from comets or meteorites that crashed into the planet, or from gases coming to the surface of the planet.

Scientists are fairly certain that ice exists on Mercury. But since we have so far only mapped about half of the planet during our one mission there, we can't be absolutely positive that the notion of some craters being in perpetual shadow is accurate. One of the goals of a future mission to Mercury is to confirm that the ice is there.

5

FUTURE MISSIONS TO MERCURY

Scientists have gleaned much from the wealth of information that the *Mariner 10* mission obtained. In order to learn more, we need to return to Mercury. The United States, Europe, and Japan all have plans for future missions to Mercury. Launch plans for the first mission back to Mercury are scheduled for 2004. Mark Robinson, one of the scientists who studied the possibility of volcanoes on Mercury, said of the next mission, "One of the things that's really neat and exciting about studying Mercury is the chance to see [the half] that's never been seen before."

NASA'S *MESSENGER*

MESSENGER stands for the **ME**rcury **S**urface, **S**pace **EN**vironment, **GE**ochemistry and **R**anging mission. It is NASA's next scheduled visit to Mercury. The mission's main goals are to study the magnetic field, atmosphere, core, mantle, geologic history, and surface of the planet. *MESSENGER* will also search for evidence of ice at the poles.

The space probe is scheduled to launch in March 2004 from Cape Canaveral, Florida. The mission will make two flybys of Venus before heading to Mercury. It should arrive at Mercury for the first flyby in

MESSENGER is NASA's next scheduled mission to Mercury. The space probe is scheduled to launch in March 2004.

July 2007. A second flyby is expected in April of the next year. Both times, it is anticipated that the probe will pass Mercury at an altitude of 124 miles (200 km). That is even closer than *Mariner 10* got to the planet. Following the flybys, *MESSENGER* will stay in orbit around Mercury for about one Earth year.

MESSENGER intends to map most of the planet that was not mapped by *Mariner 10*. It will also create a full-color map of the entire planet. There are six key questions scientists have that *MESSENGER* will set out to help them answer. All of the instruments onboard the probe were designed with these questions in mind.

Density. There have been several theories as to why Mercury is so dense. *MESSENGER* will make detailed measurements of Mercury's surface using X-ray, visible-infrared, and gamma-ray spectrometers to determine the elements and minerals that make up the surface rocks. By knowing which elements are present and in what amounts, scientists will be better able to determine which theory is correct.

Geologic history. Using the spectrometers to study the elemental and mineral makeup of different surface areas will also help scientists figure out the history of Mercury. There are areas that contain enormous, long ridges and regions thought to be volcanic. Determining the makeup of these features will give us clues as to how they were formed when Mercury was young.

Mercury's core. Since a conducting core is believed to be necessary to support a magnetic field, *Mariner 10*'s discovery of Mercury's magnetic field led to theories about the planet's core. Earth's magnetic field is generated by liquid molten iron. But because Mercury is so much smaller than Earth, scientists believe the core should have long since cooled and become solid. How could Mercury have a magnetic field without a liquid core? *MESSENGER* hopes to solve this riddle.

Atmosphere. Scientists are fairly certain that the elements of Mercury's atmosphere come from two main sources—the solar wind and surface rocks. But so far they haven't been able to determine which elements come from which source. To do this, an ultraviolet spectrometer and an energetic particle spectrometer will be used to study the chemical makeup of the atmosphere. What these spectrometers find will be compared to the results of the X-ray, gamma-ray, and visible-infrared spectrometer studies of the surface rocks from the density and geologic history experiments. Scientists will then be able to tell which elements in the atmosphere come from the solar wind and which are generated into the atmosphere from surface rocks on Mercury or from gases from the interior.

Polar ice. Earth-based radar information suggests that Mercury has pockets of ice trapped at the poles. Only a closer look will help scientists determine if that theory is correct. Instruments onboard *MESSENGER* will analyze those areas to find out if they contain hydrogen. If hydrogen is discovered, the bright areas that have been seen with radar technology are ice. *MESSENGER* will also measure the thickness of the ice.

Magnetic field. *Mariner 10* discovered that Mercury had a weak magnetic field, but we still don't know much about it. *MESSENGER* will carry a magnetometer to Mercury to determine exactly how strong the field is and how the Sun affects it. Mercury has a magnetic field that is similar to, although much smaller than, Earth's. Therefore, what we learn about Mercury's field will also expand our understanding of how Earth's magnetic field works.

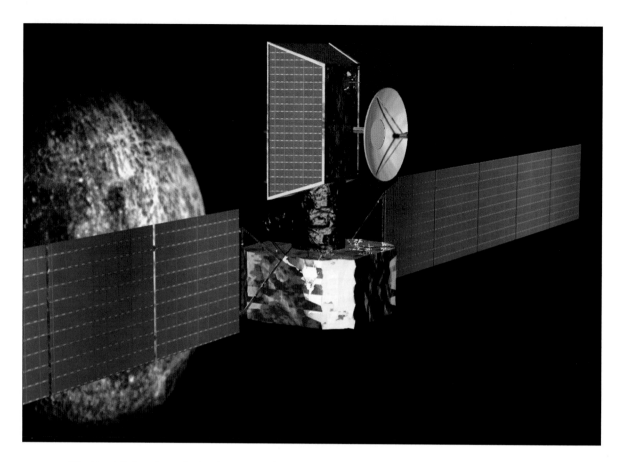

The BepiColombo *mission is a joint effort by Japan and several European countries to send a mission to Mercury. The launch is scheduled for August 2009. The probe is designed to separate into three separate spacecrafts, each of which will have a separate mission.*

JAPAN AND EUROPE WORK TOGETHER

One of the most important things to understand about space travel is that plans constantly change. Changes take place for many reasons—lack of money, technical problems, and even weather can alter the course of a mission. Traveling through space to reach another planet is such a monumental task that each detail must be as perfect as possible to achieve success.

In our attempt to learn as much as we can about other worlds, Japan and several European nations have joined forces in a remarkable effort. Japan's space agency—the Institute of Space and Astronomical Science (ISAS)—and the European Space Agency (ESA) are working together to send a probe to Mercury. The *BepiColombo* mission is scheduled to launch in August 2009.

BepiColombo is named for an Italian scientist named Guiseppe Colombo. The *BepiColombo* includes two orbiters and a lander that will be designed to separate into three spacecraft: the Mercury Planetary Orbiter (MPO), the Mercury Magnetospheric Orbiter (MMO), and the Mercury Surface Element (MSE).

The MPO is the largest of the three pieces, weighing about 1,014 pounds (460 kg). It will carry a variety of cameras and spectrometers for imaging and scientific experiments. The varying altitudes on the surface of Mercury will be measured and imaged, allowing for the first three-dimensional map of Mercury. Elements in the atmosphere will be analyzed, as well as the chemical composition of the surface. The MPO will look for ice beneath the surface. It will also study Mercury's gravity. Plans are for the MPO to orbit Mercury for one year.

Instruments on the MMO include imaging equipment, magnetometers, and charged-particle detectors. The primary goal of the MMO is to study Mercury's magnetic field and how it interacts with the solar wind. The MMO will transmit information both back to Earth and to the MSE for about a year.

The *BepiColombo* mission plans to touch down on Mercury with the MSE. The MSE is a small lander, less than 3 feet (0.9 m) across and weighing only about 66 pounds (30 kg). It will carry a camera to use on the surface and another to use while making its descent to the planet. The goals of the MSE include measuring any seismic activity, boring into the soil for sampling, and deploying a tiny robotic camera to photograph rocks and other land features. The lander is expected to function for about a week on Mercury's surface. If this project is

GUISEPPE COLOMBO

Italian scientist Guiseppe Colombo was born in 1920. He was trained as an engineer and mathematician. Colombo was a professor at the University of Padua, in northeastern Italy. This university was founded in 1222 and has a rich history. Italian astronomer Galileo Galilei also taught there. Colombo is noted for two major contributions to the field of astronomy.

In 1974 he came up with an idea that allowed special tethers, or ropes, to be used in space. Colombo helped design the Tethered Satellite System (TSS), which was used to connect a small satellite to a platform orbiting in space. The best-known application of space tethers is in keeping astronauts safely connected to their spacecraft while moving about in open space. Without the use of tethers, astronauts would not be able to perform such tasks as assembling the International Space Station from outside a craft.

Guiseppe Colombo was also a key figure in the *Mariner 10* mission to Mercury. His calculations helped make it possible for the probe to go into an orbit that would allow it to return past the planet many times. The wealth of information the mission gathered was directly related to *Mariner 10*'s ability to complete three flybys of the planet. The next mission to Mercury is named for him—the *BepiColombo*.

This artist's rendering of the BepiColombo *shows how close it will come to the hidden planet. If possible, the spacecraft will actually land on it.*

successful, *BepiColombo* will become the first mission to land on Mercury.

These future missions will all add to our wealth of knowledge. Scientists will spend many years unraveling the surprising new information about the planet closest to the Sun that these missions will uncover.

GLOSSARY

aphelion the farthest point from the Sun in an object's orbit around it

asteroid a rocky body that revolves around the Sun; a large meteoroid

comet a heavenly body that looks as if it has a long tail. It travels around the Sun and is composed of ice, frozen gases, and dust

escarpment a cliff formed by faulting or erosion that separates the surface of the land at different levels

infrared the part of the electromagnetic spectrum just beyond the red end of the visible spectrum, which is not visible to the naked eye

magnetic field the area in which a magnetic force can be felt

meteorite a solid particle that crashes into the surface of a planet

meteoroid a solid particle flying through space

perihelion the nearest point from the Sun in a planet's orbit around it

revolution the amount of time it takes a planet to orbit the Sun—a planet's year

rotation the amount of time it takes an object to spin completely around on its axis—a planet's day

satellite an object, either natural or humanmade, that revolves around a planet

sinkhole a natural depression found in areas that contain limestone

solar wind a stream of charged particles given off by the Sun that travels throughout the Solar System

ultraviolet electromagnetic radiation beyond the violet end of the visible spectrum, which is not visible to the naked eye

FIND OUT MORE

Branley, Franklyn M. *The Sun and the Solar System*. New York: Twenty-First Century Books, 1996.

Bredeson, Carmen. *Our Space Program*. Brookfield, CT: Millbrook Press, 1999.

Clay, Rebecca. *Space Travel and Exploration*. New York: Twenty-First Century Books, 1996.

Couper, Heather and Nigel Henbest. *Is Anybody Out There?* New York: Dorling Kindersley, 1998.

Redfern, Martin. *The Kingfisher Young People's Book of Space*. New York: Kingfisher, 1998.

VanCleave, Janice. *Astronomy for Every Kid*. New York: John Wiley & Sons, 1991.

Vogt, Gregory. *Scientific American Sourcebooks: The Solar System*. New York: Twenty-First Century Books, 1995.

———. *Mercury, Venus, Earth, and Mars*. New York: Raintree Steck-Vaughn, 2001.

BIBLIOGRAPHY

Cowen, Ron. "Mercury: A Forgotten Planet." *Science News*. July 8, 2000.

Dunne, James A. "Mariner 10 Mercury Encounter." *Science*. July 12, 1974.

Fradin, Dennis Brindell. *The Planet Hunters*. New York: Margaret McElderry Books, 1997.

Henbest, Nigel. *The Planets: A Guided Tour of our Solar System Through the Eyes of America's Space Probes*. New York: Viking, 1992.

Hooke, Adrian J. "The 1973 Mariner Mission to Venus and Mercury, part one." *Spaceflight*. January 1974.

————. "The 1973 Mariner Mission to Venus and Mercury, part two." *Spaceflight*. February 1974.

Isbell, Douglas. "NASA Selects Missions to Mercury and a Comet's Interior as Next Discovery Flights." NASA Press Release, July 7, 1999.

Miller, Ron and William K. Hartmann. *The Grand Tour: A Traveler's Guide to the Solar System*. New York: Workman Press, 1981.

Murray, Bruce and Eric Burgess. *Flight to Mercury*. New York: Columbia University Press, 1977.

Murray, Bruce. *Introduction to the Planets: Readings from Scientific American*. San Francisco, CA: W. H. Freeman and Company, 1983.

Nourse, Alan E. *Nine Planets*. New York: Harper & Row, 1970.

Pasachoff, Jay M. and Donald H. Menzel. *A Field Guide to the Stars and Planets*. New York: Houghton Mifflin, 1992.

Robinson, Mark S. and Paul G. Lucey. "Recalibrated *Mariner 10* Color Mosaics: Implications for Mercurian Volcanism." *Science*, vol. 275, 1997, pp. 197–200.

Sagan, Carl. *Cosmos*. New York: Random House, 1980.

_____. *Pale Blue Dot: A Vision of the Human Future in Space*. New York: Random House, 1994.

Sheehan, William. *Worlds in the Sky: Planetary Discovery from Earliest Times Through Voyager and Magellan*. Tucson, AZ: The University of Arizona Press, 1992.

Watters, Thomas R. *Planets: A Smithsonian Guide*. New York: Macmillan, 1995.

WEBSITES

The following websites are all recommended to learn more about Mercury:

http://www.solarviews.com/eng/mercury.htm
This site contains all kinds of information on Mercury, including statistics and an exploration timeline. It also includes many photos, maps, and even animations of the planet's formation.

http://www.seds.org/billa/tnp/mercury.html
This government-sponsored site gives basic information on Mercury, but also provides links to many other sites, which contain photographic images and more specific information on what was learned from the *Mariner 10* mission.

http://sd-www.jhuapl.edu/MESSENGER
This NASA site is full of information about the proposed *MESSENGER* mission to Mars, scheduled for 2004.

http://nssdc.gsfc.nasa.gov/planetary/ice/ice_mercury.html
This NASA-sponsored site explains why scientists believe there is ice on Mercury.

ABOUT THE AUTHOR

Tanya Lee Stone is a former editor of children's books who now writes full time. She holds a master's degree in science education and is the author of more than twenty books for young readers, including *Saturn, Rosie O'Donnell: America's Favorite Grownup Kid*, and *The Great Depression and World War II*. She lives in Vermont with her husband, Alan, and their two children.

INDEX

Page numbers for illustrations are in bold.

atmosphere, 26, 28, **29**, 30, **30**, 31, 48, 51, 53

core, **34**, 35, 44, 48, 50–51

craters, 23, 36, **37**, **38**, 39–40, 44, **45**, **46**, 47

distance from Earth, 18

distance from Sun, 26, 32

first dated observations, 9

future missions, 35, 47, 48–53, **49**, **52, 55**

gravitational field, 32

ice, 47, 48, 51, 53

magnetic field, 24–26, 28, 35, 48, 50–51, 53

map, 9, 12, 16, 23, **31**, 47, 50, 53

mass, 9, 10, 23, 32

orbit, 9, 10–11, 32, **33**

period of revolution (year), 18, 20

period of rotation (day), 18, 20, 24, 26, 32, 47

phases, 9

place in the Solar System, 9, 17, 18, 47, 55

plains, 23, 39–40, **41**, 44, **45**

poles, 32, **46**, 47, 48, 51

scarps, 44, **45**

shape, 23

size, 26

temperature, 26, 28, 31–32

viewed from Earth, **7**

viewed from space, **8**, **29**, **46**, **49**

Mercury Magnetospheric Orbiter (MMO), 53

Mercury Planetary Orbiter (MPO), 53

Mercury Surface Element (MSE), 53

MESSENGER, 48, **49**

mystery planet. *See* Vulcan

National Aeronautics and Space Administration (NASA), 20–21, 43, 48

Neptune, 42

Newton, Isaac, 11

outer space. *See* space

Pioneer, 21

planets, 6, 9, 11, 21, 31. *See also* individual planets

plasma-measuring experiment, 23, 27

Pluto, 9, 42

precession, 11

Robinson, Mark, 44, 48

satellites, 31

Saturn, 42

Schiaparelli, Giovanni, 9, 12, 16, **17**

Science, 44

Solar System, 9, **16**, 20, 31, 36, 39

solar wind, 20, 23–24, 27, 28, 51, 53

space, 24–25, 27, 28, 30, 52, 54

stars, 6, 43

Stratospheric Observatory for Infrared Astronomy (SOFIA), 43

Sun, 9, 10, 17, 18, 20–21, 23, 25–27, 31–32, 47, 51

telescopes, 6, 9, **12**, 13–15, 23, 27, 43
  astronomical, 15
  radio, 18, **19**, 20, 24–25, 47

Television Photography experiment, 23

Tethered Satellite System (TSS), 54

Timocharis, 9

Two-Channel Infrared Radiometer experiment, 26

U.S. Geological Survey, 44

Uranus, 42

Venus, 20, 39, 41, 48

Very Large Array (VLA) telescope. *See* radio telescope

Voyager, 21

Vulcan, 10

Zupus, Giovanni, 9